Jeremy Griffith's Homage To Steve van Hemert

Jeremy Griffith

Please read Anthony Gowing's important Introduction before watching the video of Jeremy's speech at www.humancondition.com/jeremy-griffith-homage-to-steve-van-hemert

OR

Scan code to view

Jeremy Griffith's Homage To Steve van Hemert by Jeremy Griffith

Published in 2026 by WTM Publishing and Communications Pty Ltd (ACN 103 136 778) (www.wtmpublishing.com).

All enquiries to:
FIX THE WORLD®
Email: info@humancondition.com
Website: www.humancondition.com

FIX THE WORLD (FTW) is a global not-for-profit movement represented by FTW charities and centres around the world.

ISBN 978-1-74129-118-6
CIP – Biology, Philosophy, Psychology, Health

Filming by Tess Watson; editing by James Press.

Commendations for Griffith's treatise

From Thought Leaders

'[**Prof. Stephen Hawking**] is most interested in your impressive proposal.' ● 'In all of written history there are only 2 or 3 people who've been able to think on this scale about the human condition.' **Prof. Anthony Barnett**, zoologist ● '*FREEDOM* is the book that saves the world...cometh the hour, cometh the man.' **Prof. Harry Prosen**, former Pres. Canadian Psychiatric Assn. ● '*FREEDOM* is a really significant contribution to science, providing fundamental answers.' **Prof. Alejo Vidal-Quadras**, former Vice-Pres. of the European Parliament & nuclear physicist ● '*FREEDOM* is the rarest gift...it makes sense of the human condition...tells the truth...is next only to the Bible...ends the moral corruption across the world.' **Prof. Yonas Adaye**, Addas Ababa Uni. & Commissioner, Ethiopian Dialogue Comm. ● 'I am stunned and honored to have lived to see the coming of "Darwin II".' **Prof. Stuart Hurlbert**, esteemed ecologist ● 'Living without this understanding is like living back in the stone age, that's how massive the change it brings is!' **Prof. Karen Riley**, clinical pharmacist ● 'Frankly, I am blown away by the ground-breaking significance of this work.' **Prof. Patricia Glazebrook**, philosopher ● 'I've no doubt a fascinating television series could be made based upon this.' **Sir David Attenborough** ● '*FREEDOM* is the necessary breakthrough in the critical issue of needing to understand ourselves.' **Prof. David J. Chivers**, former Pres. Primate Society of Britain ● 'Whack! Wham! I was converted by Griffith's erudite explanation for our behaviour.' **Macushla O'Loan**, *Executive Women's Report* ● 'This is indeed impressive.' **Dr Roger Lewin**, preeminent science writer ● 'I have recommended Griffith's work for his razor-sharp biological clarifications.' **Prof. Scott Churchill**, psychologist ● 'An original and inspiring understanding of us.' **Prof. Charles Birch**, zoologist ● 'The insights are fascinating and pertinent and must be disseminated.' **Dr George Schaller**, pre-eminent biologist ● 'Very impressive, particularly liked the primatology section.' **Prof. Stephen Oppenheimer**, geneticist, author *Out of Eden* ● 'I consider the book to be the work of a prophet.' **Dr Ron Strahan**, former dir. Sydney Taronga Zoo ● 'The scholarly value is comparable to several of the most celebrated publications in biology.' **Prof. Walter Hartwig**, anthropologist ● 'You're on to getting answers to much that has bewildered humans.' **Dr Ian Player**, famous Sth. Afr. conservationist ● 'A superb book, a forward view of a world of humans no longer in naked competition.' **Prof. John Morton**, zoologist ● 'This might bring about a paradigm shift in the self-image of humanity.' **Prof. Mihaly Csikszentmihalyi**, psychologist ● 'FTW is an island of sanity in a sea of madness.' **Tim Macartney-Snape**, world-leading mountaineer & twice Order of Australia recipient

Commendations From The General Public

'Griffith should be given Nobel prizes for peace, biology, medicine; actually every Nobel prize there is!' ● 'He nailed it, nailed the whole thing, just like the world going from FLAT to ROUND, BOOM the WHOLE WORLD CHANGES, no joke.' ● '*FREEDOM* will be the most influential, world-changing book in history, and time will now be delineated as BG, before Griffith, or AG, after Griffith.' ● 'I'm speechless – this is bigger than natural selection & the theory of relativity!' ● 'I really think this man will become recognized as the best thinker this world's ever seen, and don't we need him right now!' ● 'Griffith has decoded the human species, we FINALLY know what's going on & the suffering stops!' ● 'The world can't deny this for much longer, let the light in, save the human race!' ● 'This is the most exciting moment in my life. *THE Interview* tore my hat off & let my brain fly into the sky!' ● '*THE Interview* should be globally broadcast daily. The healing explanation humans so sorely need.' ● 'In a world that's lost its way there's no greater breakthrough, water to a world dying of thirst.' ● 'Dawn has come at Midnight! A brilliant exposition, we could be on the cusp of regaining Paradise!' ● 'This man has broken the great silence, defeated our denial, got the truth up, woken us from a great trance.' ● 'Beware the 'deaf effect; your mind will initially resist the issue of our corrupted condition and so find it hard to take in or hear what's being said, but if you're patient you'll find the redeeming explanation of our condition pure relief.' ● 'John Lennon pleaded "just give me some truth", well this site finally gives us *all* the truth!' ● '*FREEDOM* is the most profound book since the Bible, now with the redeeming truth about us humans.' ● '*Death by Dogma* is brilliant clarification.' ● 'We were given a computer brain, but no program for it; but Aha, Griffith has found it, made sense of our lives!' ● 'This just goes deeper & deeper in explaining us, like dawn devouring darkness, amazing!' ● 'Agree, this is not another deluded, pseudo idealistic, PC, 'woke', false start to a better world, but the human-condition-resolved real solution.' ● 'Freedom indeed! What we have here is the second coming of innocence who exposes us but sets us free!' ● 'As prophesised, King Arthur has returned to save us (mentioned in par.1036 *Freedom*)' ● 'We all need to go back to school & learn this truthful explanation of life.' ● 'Join in our jubilation, your magic reunites, all men become brothers, all good all bad, be embraced millions! This kiss [of understanding] for the whole world' – From Beethoven's 9th (par.1049 *Freedom*)

Contents

	PAGE	PARA-GRAPH
Background	6	
Introduction, by a FIX THE WORLD Director, Anthony Gowing	9	1–
Transcript of Jeremy Griffith's homage to Steve van Hemert	15	11–
Postscript: Comments about Jeremy's speech on FIX THE WORLD's Facebook Group	29	45–

Background

Jeremy Griffith is an Australian biologist who has dedicated his life to bringing redeeming and psychologically healing biological understanding to the dilemma of the human condition—which is the underlying issue in all human life of our species' extraordinary capacity for what has been called 'good' and 'evil'.

Jeremy has published over ten books on the human condition, including:

— *Beyond The Human Condition* (1991), his widely acclaimed second book;

— *A Species In Denial* (2003), an Australasian bestseller;

— *FREEDOM: The End Of The Human Condition* (2016), his definitive treatise;

— *THE Interview* (2020), the transcript of acclaimed British actor and broadcaster Craig Conway's world-changing and world-saving interview with Jeremy about his book *FREEDOM*;

— *Death by Dogma: The biological reason why the Left is leading us to extinction, and the solution* (2021), which presents the biological reason why Critical Theory threatens to destroy the human race;

— *The Great Guilt that causes the Deaf Effect* (2022), which describes how lifting the great burden of guilt from the human race initially causes a 'Deaf Effect' difficulty taking in or 'hearing' what's being presented;

— *The Shock Of Change that understanding the human condition brings* (2022), which addresses how to manage the shock of change that inevitably occurs when the redeeming understanding of our corrupted condition arrives;

— *Therapy For The Human Condition* (2023), which is about the therapy that is desperately needed to rehabilitate the human race from our psychologically upset state or condition, elaborating on what is presented in *FREEDOM*;

— *Our Meaning* (2023), which explains how being able to know and fulfil the great objective and meaning of human existence finally ends human suffering;

— *The Great Transformation: How understanding the human condition actually transforms the human race* (2023), which gives a concise description of how the psychological rehabilitation of humans occurs, and how everyone's life can immediately be transformed; and

— *AI, Aliens & Conspiracies: The Truthful Analysis* (2023), which provides Jeremy's thoughts on the much discussed question of the danger of Artificial Intelligence (AI), and on the possibility of alien life visiting Earth, and also his explanation for the epidemic of conspiracy theories.

— *Sermon On The Beach* (2024), which is Jeremy's elaborated transcript of his inspired description of how the human race now leaves the horror of the human condition forever!

— *The Human Condition: What exactly is it, what caused it, and how the human race has finally liberated itself from the horror of it* (2025), which is one of Jeremy's three most important presentations on the human condition, serving as a powerful short-in-length bridge between *THE Interview* and *FREEDOM*.

This booklet, ***Jeremy Griffith's Homage To Steve van Hemert***, is the transcript of a speech Jeremy gave about how his great friend Steve's larger-than-life, human-condition-demolishing escapades and honest humour helped Jeremy hold onto his truthful thinking and go on to solve the human condition. The video can be viewed at www.humancondition.com/jeremy-griffith-homage-to-steve-van-hemert.

Jeremy's work has attracted the support of such eminent scientists as the former President of the Canadian Psychiatric Association Professor Harry Prosen, the esteemed American ecologist Professor Stuart Hurlbert, the Spanish nuclear physicist and former Vice-President of the European Parliament Professor Alejo Vidal-Quadras, Australia's Templeton Prize-winning biologist Professor Charles Birch, the former President of the Primate Society of Great Britain Professor David Chivers, world-leading physicist Stephen Hawking, as well as other distinguished thinkers such as the pre-eminent philosopher Sir Laurens van der Post.

Jeremy is the founder and a patron of FIX THE WORLD (formerly World Transformation Movement)—see www.HumanCondition.com.

With the real problem of the human condition finally solved we can now ACTUALLY fix the world!

Introduction, by a FIX THE WORLD Director, Anthony (Tony) Gowing

[1]This is gold. It's Jeremy doing his thing on 21 March 2026—taking the false world head on, standing up to the artificial and superficial, sophisticated, all-confident and all-dominant, 'There's nothing wrong with our world or us, don't embarrass us by confronting us with the truth of our corrupted condition' attitude that exists everywhere in the world. 'Just leave us alone Jeremy so we can get on with our resigned lives where we promenade about as got-it-together legends' when, as Jeremy compassionately explains in all his books, we are actually all, the whole human race, in a dire state of terminal alienation!

[2]This video and its transcript is of Jeremy giving a speech at his very good friend Steve van Hemert's surprise 80th birthday party where many close friends of Steve's had gathered to celebrate the contribution he had made to their lives. Jeremy is glorifying Steve's truth-telling honesty and how that honesty helped Jeremy go on to solve the human condition, find the redeeming and liberating good reason for why we humans have suffered from the corrupted state of the human condition. You will see that Jeremy was boldly making it completely clear to everyone how important Steve's honesty has been to the world by creating a new Australian flag with Steve at the centre of it in the famous Australian outlaw Ned Kelly's helmet. Bold Ned Kelly, bold Steve van Hemert, and bold Jeremy Griffith!

[3]Most significantly, and interestingly, this bold spirit of defiance of all the dishonest denial of our corrupted human condition, which

Ned Kelly, Steve and Jeremy have exhibited, is a spirit that lies at the very heart of Australia's convict-colonised character—which Jeremy explains in his 2003 bestselling book *A Species In Denial* in the section 'Australia's role in the world', and also in his main book *FREEDOM* in chapter 9:11 that is titled 'The pathway of the sun'. So it makes complete sense that this defiance of all the dishonest denial in the world that led to the finding of the human-race-saving understanding of the human condition has occurred in Australia.

[4]So thank you Steve for helping Jeremy hold onto the truth about our corrupted human condition, which enabled him to continue to think truthfully enough to, as I say, go on and find the human-race-saving, redeeming biological understanding of that horrifically corrupted state that will fix the world. And thank you Steve & Robbie van Hemert, Tom & Molly Shannon (Steve & Robbie's daughter), and Manus McFadyen (a great friend of theirs), for giving Jeremy the opportunity to stand up for his true world against the dishonest false world where everyone everywhere in the world is living in a resigned state of dishonest denial of our corrupted human condition. And thank you for enabling Jeremy to do it in the middle of the Hunter Valley, which from Armidale to the coast has in the all-too-recent past been a hot-bed of outrageous lies that FIX THE WORLD is a sinister organisation—terrible lies that we finally defeated after an extremely agonising but successful 15-year-long defamation court action. And, by the way, almost singlehandedly, Tim Watson and his wife Ali, who drove Jeremy up from Sydney to this Hunter Valley gathering, stood like a colossus against all the persecution of us in that valley by courageously publicly supporting us time and again through those terrible years from their home in the middle of that valley.

[5]This speech is wonderful inspiration for us all at FIX THE WORLD. Historically it has always been important to stand up for truth, but the levels of devastation in the world now are so great

that standing up for this particular truth of Jeremy's redeeming explanation of the human condition is absolutely critical—because it is the ONLY way to save the human race. As Jeremy said to me on his return to Sydney, **'What is going to happen in the world over the next 10 or possibly more years is that more and more people are going to accept that the human race is in a completely devastated and desperate situation, but that there is a way out of this horrific predicament, and that is to support this redeeming and psychologically healing scientific explanation of the human condition. More and more people are going to realise that you can set out to do 'this', or strive to achieve 'that', but the fact is it will make no difference to the dire situation the world is in—ONLY SUPPORTING THIS INITIALLY-CONFRONTING-BUT-ULTIMATELY-RELIEVING EXPLANATION OF THE HUMAN CONDITION CAN SAVE OUR SPECIES. Basically the world is now in a life-or-death war where the human-race-liberating truth has to win out over the entrenched lies, the entrenched denial of our corrupted human condition. That is the predicament the world and every human in it is now in. So our job at FIX THE WORLD is to defend the human-race-liberating understanding of ourselves everywhere and anywhere—hold the fort until the cavalry of responsible establishment support arrives.'**

[6]Which is precisely what Jeremy is showing us how to do in this video—how to defy all the denial, all the silence, all the darkness that wants to resist the human-race-saving understanding we now have of our corrupted human condition.

[7]Yes, anyone can figure this out for themselves once they understand this information—that the world is mad and rapidly going completely crazy, but there is actually a way out, and that is to support this truthful and compassionate understanding of the human condition. There is no other meaningful path forward that anyone can follow, and that is the truth! So all we at FIX THE WORLD have to do is, as Jeremy said, look after this biological understanding of the human condition until everyone turns up to help us.

Professor Yonas Adaye Adeto

[8] As evidence of this growing appreciation of the importance of Jeremy's work and thus of the importance of "holding the fort until the cavalry arrives", this is what Professor Yonas Adaye Adeto, the Director of the Institute for Peace and Security Studies at Addis Ababa University, and a Commissioner of the Ethiopian National Dialogue Commission, sent us just last week: **'I consider *FREEDOM* by Jeremy Griffith to be the rarest gift on the planet. Jeremy tells the truth and is our guide to ending the human condition. In my assessment, his books are next only to the Bible, they truly make sense and are absolutely thrilling, liberating and transformative. I have given a copy of *FREEDOM* to dozens of my friends and colleagues, including at the Ethiopian National Dialogue Commission who ended up buying and**

recommending the book to many. What else should be disseminated more than good news such as this? It does not preach a new religion; it does not degrade your colour, creed or class; it embraces humanity as they are; it ignites imagination; it empowers you to release your own potential to be kind to yourself, to your loved ones and to the world. It reinforces your religion, makes you accept yourself, and fosters respect for the rest of humanity! Most importantly, it makes sense of and ends the moral corruption we are all facing across the world today. I am glad to cite the works in my research and presentations. I just recommended that some of my PhD students keep the books as self-help readings in their bags and on their shelves wherever they go.'

Professor Alejo Vidal-Quadras

[9]And this is what Professor Alejo Vidal-Quadras, a nuclear physicist and former Vice-President of the European Parliament no less, said only a few weeks ago, **'*FREEDOM* is a really significant contribution to science, providing fundamental answers'**, and **'Jeremy's interview with Craig Conway** [which can be viewed free of charge at www.humancondition.com] **deserves spreading to the largest number of people possible.'**

[10] So watch Jeremy in the video below (or read the following transcript) of his speech in the Hunter Valley show us the only way to save the human race—how to stand up for the light of this liberating understanding against all the darkness of the entrenched denial that basically wants Jeremy and his human-race-saving truth to go away and let the human race die in a horrific state of terminal psychosis!

After reading the above Introduction, watch the video of Jeremy's speech at www.humancondition.com/ jeremy-griffith-homage-to-steve-van-hemert.

OR

Scan code to view

Transcript of Jeremy Griffith's homage to Steve van Hemert

11 **Robbie van Hemert**: So I'll just put you over to Jeremy. Good on you Jeremy. Thanks for coming up.

12 **Jeremy Griffith**: Molly [Shannon, the previous speaker], you are an absolute angel. That was beautiful, and so lovely to hear Steve's backstory.

13 What I want to say is very important to me so I have written it out. So I'm sorry about that, but it'll only take 10 minutes. This is pretty profound stuff that's incredibly important, which I'm hoping you'll get to understand in a minute.

Steve van Hemert

Jeremy Griffith

[14]I'm here to pay homage to one of the most unsung heroes in Australia, indeed *in the world*—our Steven van Hemert. In fact, to correct this appalling situation and sing his much deserved praise—[butcher bird sings in background] see the butcher birds are on side too—I've designed this new flag for Australia that features Stevie Pops in [the famous Australian outlaw] Ned Kelly's helmet—so this is a present for you Stevie. The text which I'll make sense of shortly, reads **'Our new Australian flag — our Stevie 'Popsie' man-mountain who stood up to and exposed bullshit like our Bold Ned Kelly — and laid the foundations for a new truthful world!'**

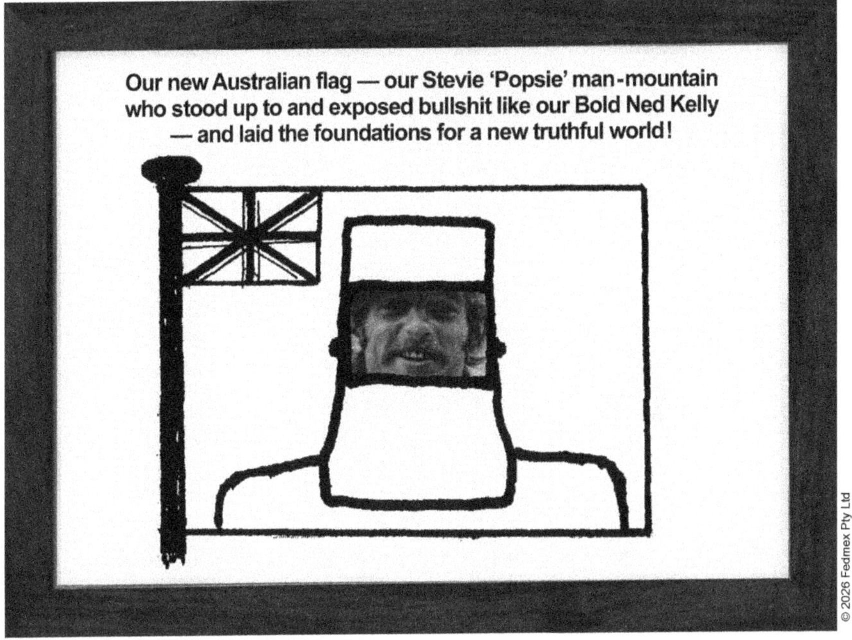

15 I've made this flag because Steve is an absolutely outstanding example of one of those super precious and very rare **'hell raisers'** who are often denigrated for their honesty—one of those **'larger than life'**, **'rebellious'**, **'high spirited'**, **'non-conformist'**, immensely entertaining and relieving but often extremely confronting demolishers of all the bullshit falseness that we have to put up with in the world. An **'outlaw'**, like Ned Kelly, who stands defiantly up against the oppressively dishonest, artificial and superficial establishment—someone who tears the scab off all the crap we have to stomach! Zorba the Greek and the hell-raisers Richard Harris, Peter O'Toole and Oliver Reed were famous examples of this high-spirited, larger-than-life character.

[16]They just don't come any better than Stevie when it comes to being a larger-than-life exposer and demolisher of all the bullshit. We have all here I'm sure greatly enjoyed and benefited from Steve's endless humour and jokes that are all based on remorseless, outrageous, irreverent, defiant honesty. God I love—we all love—Stevie van Hemert!

[17]I have just turned 80, like Stevie does this year, and we first met in our 20s back in the early 1970s at the 'Four In Hand' hotel in Paddington in the heart of Sydney where Steve was on a bender of absolute bullshit-exposing-and-demolishing escapades! Deb McDonald had brought Steve to Paddington because she thought he was so extraordinarily special, and he certainly was. He had bleached blond hair and a big upper torso from weight-lifting or something, so much so we called him 'Arms', but all that ahead-of-his-time, put-it-out-there presence wasn't what made him so special, it was the outrageously disarming honesty of his humour and wild, **'hell-raising'**, nothing-is-sacred carryings on.

[18] — I remember at the Four In Hand, John Walton, one of the young business Turks of Sydney coming in in his suit and tie, and Steve putting his arm around him and telling him what a legend he was and then 'accidentally' undoing his briefcase so all the papers fell out on the floor and then 'accidentally' stepping all over them and spilling his beer on them while 'helping' to pick them up!

[19] — On another occasion, there was this bloke who thought he was God's gift to women and Steve wrote a message on a drink coaster, seemingly from the barmaid, saying "I just think you're so special, love Suzie", and slipped it in front of him when he wasn't looking, and the whole pub was in on it, watching this bloke whispering to the barmaid, who had no idea what he was talking about, that they should meet up later!

[20] — And this sort of stuff would go on all the time, right up to this day, which I'm sure everyone here who knows Steve has joyously

experienced. We actually tried to get Steve on radio because he was so entertaining but his honesty was just too deadly for the world, he was just fucking right out there!

[21] — On the weekends we would go on adventures, like out to the Polo at Windsor, and Steve would entertain everybody, like standing in one of the 44 gallon garbage bins in front of everybody, saying "Lord & Lady Muck and all you other distinguished guests", and ramble on with all manner of bullshit about their distinguished world! Honestly they loved it.

[22] — One year we were invited to the annual Black & White ball at the Opera House, and I remember Stevie had some sort of friendly liaison with the Matron of Honour under the tableclothed main, official table in the ballroom! Needless to say, that was the last time we were invited to that event!

[23] — If anybody invited us to a party, we would turn up with tampons in our nose and ears, you name it, we were crazy!

[24] — In the end, no one would invite us to anything—except a few like one-armed Pete Mansfield, and yeah Rosie at her restaurant in Oxford St Paddington where we would go when the pubs closed to get a feed and say hello to her two pretty daughters. Despite causing chaos every time we visited, Rosie never kicked us out. I remember one time Steve entertaining a table of diners who were celebrating a birthday, and Steve got up on the table and stood right in the middle of the birthday cake and dolloped bits of it on everyone around the table! Talk about tear the crap off all the artificial, pretentious bullshit with humour, Stevie is just amazing!

[25] — And I should say that our very, very good friends here Manus McFadyen and Tim Watson took part in some of these wild escapades over the years, so they were part of all this high-spirited wickedry!

L to R: Tim Watson, Steve van Hemert, Ali Watson, Molly Shannon, Robbie van Hemert, Manus McFadyen, Tom Shannon

26 I will give you even more dramatic examples of these carryings on shortly, but firstly I'll very briefly explain what is meant by the term **'larger than life'** that describes Stevie.

27 I'm a biologist who has written over 20 books about the human condition, which is the issue of why we humans are the way we are, such deludedly artificial and superficial beings, so I'm in some sort of a position to explain what we mean by **'larger than life'**.

28 You can read about this in my books, but one of the greatest of all philosophers, Plato, in his greatest work, *The Republic*—and this is the central part of *The Republic*, so it's one of the greatest bits of writing ever—truthfully described humans as being **'prisoners'** in a **'dark cave'** of extremely dishonest denial, fearfully hiding from the truth of their corrupted **'human condition'**. So you've got to envisage this. Alfred North Whitehead, one of the greatest philosophers of the 20th century, described the history of philosophy (philosophy being the study of **'the truths underlying all reality'**) as being merely **'a series of footnotes to Plato'**. So Plato was *way out there* in front of everybody in terms of his ability to talk truthfully about the human condition, why we are the way we are. And so what Plato had to say—and this is his central work, *The Republic*—truthfully

describes humans as being **'prisoners'**—that's his term—in a **'dark cave'** of extremely dishonest denial, fearfully hiding from the truth of their corrupted **'human condition'**.

29 So, in terms of the human condition, **'larger than life'** characters like Steve break the spell of all this dishonest cave-dwelling denial, tear the scab off all the crap, and by so doing bring deep relief from all the deluded dishonesty going on everywhere.

30 So you can imagine how much relief it was for me, having had a very sheltered upbringing and as a result being so distressed and perplexed by all the dishonesty in the world, to run into Steve who tore bullshit apart all day long—so we became very great friends. It was that honesty and truthful thinking that helped me to go on and find the understanding of the human condition that saves the human race, as indicated in my Steve van Hemert flag. Because it says here, it **'laid the foundations for a new truthful world.'** So I'm talking really big stuff here, really the most significant things that are happening in the world today, *the most significant thing* happening in the world today. To quote Professor Harry Prosen, former President of the Canadian Psychiatric Association—so this is one of the leading psychiatrists in the world: **'I have no doubt that Jeremy's biological explanation of the human condition is the holy grail of insight we humans have sought for the psychological rehabilitation of the human race.'** And don't we need that *right now*, with all the madness happening everywhere.

31 So now I want to tell you about two particularly wild adventures we went on. The bushrangers Ned Kelly and Ben Hall were outrageously-defiant-of-the-dishonest-establishment **'outlaws'**, who regularly held up whole towns, locking everyone in a pub and having one big party. In the case of Ben Hall, in 1863 his gang held up the whole of Bathurst which is only 150 kilometres from here. Well, Steve and I once took over Sydney, dressed up as the outlaws Butch Cassidy and the Sundance Kid (which Molly has a famous photo of that a photojournalist took, shown overleaf).

[32]In the photo you can see the Lord Mayor of Sydney, Nick Shehadie, that's Stevie, that's me, we're dressed up as Butch Cassidy and the Sundance Kid, and the girl in the right-hand corner is Dolly from [the sitcom] *Number 96*. We've got these toy pistols. This is the official stage, we were as pissed as newts, and so we took over that stage and we've been told since we were lucky we weren't shot by the police. So in the photo you can see the Lord Mayor of Sydney, Nick Shehadie, and his entourage of celebrities that we are holding up with toy pistols at the annual Rocks Festival in Sydney in 1975; a day of infamy where we ended up in jail—this despite the crowd that had been following us and growing all day trying to tip the paddy wagon over to free us—which was all reported in the newspaper the next day!

[33] On another occasion we went to Melbourne and took over the famous 'Tok H' (Toorak Hotel) in the heart of Melbourne and so took that town over, and then went down to the wharfs in Melbourne determined to jump a ship and go to London and take that joint over too, before we finally ran out of steam.

[34] But before running out of puff, I remember thinking the early Vikings adventured all the way to America—well, I was thinking we wouldn't have stopped there, we would have gone all the way down the coast of America and then around the bottom and across to Tahiti where all the pretty girls are, and then to the moon, why not, I mean why fucking not, the world is so exciting—and in huge need of having all its bullshit exposed by the likes of Stevie!

[35] I could tell you the unbelievable details of those two adventures to the Rocks Festival and to Melbourne, but I think everyone's got the idea about the magnificence of Stevie boy. [There is a four-hour 18 November 2024 video, which can be viewed at www.humancondition.com/deductive-inductive-science-talk, where, amongst other things, Jeremy describes Steve's and his high-spirited 'take over' of Sydney and Melbourne.]

[36] So thank you Stevie from the bottom of my heart, and from the whole world, for all your fabulously courageous honesty. And very importantly, because you have lived such a defiant life, all the

love in the world to Robbie, Molly and Manus, and now Tom, for looking after you and keeping your defiant self alive all these years. And let's hope stem cell research comes up with a fix for Steve's Parkinson's Disease.

37 So yeah, thank you Stevie from the bottom of my heart, mate. [Applause]

[38]That's one of my main books called *FREEDOM*. This other booklet is the transcript of an interview I've done which has gone viral around the world. And we've got over 80 Centres around the world now supporting this stuff, and we've got over 100,000 Facebook Group members and as I said, there's 20 books, and I'm working on another book.

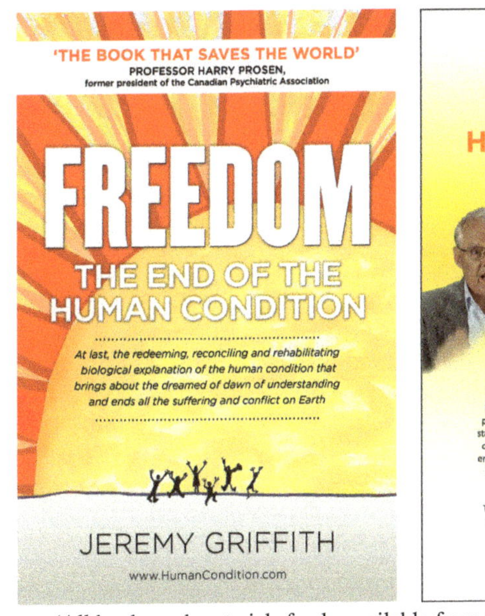

(All books and materials freely available from www.humancondition.com.)

[39]This is a good book, it's one I've just finished. It's called *The Human Condition* (overleaf). And it says on the front cover, **'What exactly is it, what caused it, and how the human race has finally liberated itself from the horror of it'**.

THE HUMAN CONDITION

What exactly is it, what caused it, and how the human race has finally liberated itself from the horror of it

Our present dystopia caused by our fear of the scorching 'fire' of the issue of the human condition, and the quenching of that 'fire' and resulting liberation of humanity from the agony and horror of the human condition

Jeremy Griffith

www.HumanCondition.com

40 And I've drawn this picture of this collapsing world that says **'science'** and **'human progress'** all disintegrating, people collapsed, vegetation collapsed. And then there's this fire, this terrifying fire, which is the subject of the human condition which no one can go near; everyone's living in fearful denial of this terrifying subject. As Plato said, they're living in this **'cave'**, can't go near the truth of who they really are because it's unconfrontable, it's unbearable. But *one day, the human race has lived in hope, faith and trust that we would find the redeeming understanding of our corrupted condition*. And then and only then could we liberate ourselves from the horror of the human condition and become sound and secure again. So [then I've depicted] putting out that fire so that we can all go to this human-condition-free state. And we're getting commendations now from a broad range of eminent scientists, thought leaders and members of the general public from around the world—you can see all that on the Commendations & Reviews page on our website at www.humancondition.com/reviews-commendations.

41 So we have these flyers that people are handing out around the world.

42 This is a summary of the commendations from scientists. These are some of the leading scientists in the world, Professor Stephen Hawking, this is Professor Anthony Barnett, a zoologist who had a science show in Canberra, saying, **'In all of written history, there are only 2 or 3 people who've been able to think on this scale about the**

human condition'. It's a very, very difficult and dark subject to face, which is why this project of bringing understanding to the human condition was viciously attacked, and we had to endure that. All scientific new ideas are subject to initial condemnation and have to survive that. So we're through that stage. We're now getting lots and lots of support. This is a quote that's come in this week: the former Vice-President of the European Parliament—you can't get much higher up than that—and nuclear physicist, Professor Alejo Vidal-Quadras, said, **'*FREEDOM* is a really significant contribution to science, providing fundamental answers'**, and my now famous interview with Craig Conway, **'deserves spreading to the largest number of people possible'** [this interview can viewed at the top of our website www.humancondition.com].

[43] Steve's daughter Molly's been a great supporter of this information.

Molly Shannon; L to R: Annie Williams, Tom Shannon, Fergus Shannon, Jeremy Griffith, Molly Shannon

[44] So the human condition *is* a difficult subject. And the ability to think truthfully about this forbidden subject is in large measure due to Stevie. So I just want to pass that on to everybody. [Applause]

Postscript:
Comments about Jeremy's speech on FIX THE WORLD's Facebook Group

[45] Annie Williams: As Tony said, thank you very much Steve & Robbie, Tom & Molly, and Manny, for making possible this powerful presentation by Jeremy of his human-condition-understood truthful world in the heart of the Hunter Valley, which was one of the places during the 1990s where there was horrible persecution of Jeremy, his supporters and his work, which thank goodness we had the strength to successfully defeat through a 15-year-long defamation court action.

[46] Two quotes I think are particularly relevant.

[47] The playwright George Bernard Shaw saying **'All great truths begin as blasphemies'**, with the greatest blasphemy of all being the truthful explanation that Jeremy delivers of our universally denied corrupted human condition—hence the resistance to Jeremy's human-race-saving work that we at FIX THE WORLD had to, and did, overcome.

[48] And with regard to the courage of our defiance of that persecution, this quote by the science historian Thomas Kuhn is particularly relevant: **'In science...ideas do not change simply because new facts win out over outmoded ones...Since the facts can't speak for themselves, it is their human advocates who win or lose the day.'**

[49] Susan Armstrong: How magnificent! Straight down the middle! It's so courageous it brings me to my knees showing us **'the only way to save the human race—how to stand up for the light of this liberating understanding against all the darkness of the entrenched denial'**. Let's go, let's **'kick the darkness until it bleeds daylight'** (as U2 sang) all over the globe!!

[50] Connor FitzGerald: Awesome video. Standing up for the Truth against the silent world of denial is so bloody important it's not funny. So good to see those commendations from Professor Adaye and Professor Vidal-Quadras too!!

[51] Gerry St Onge: So great to see Jeremy again speaking on this incredible subject, and to get insight into his youth and defiance of this crazy train we've been on racing toward terminal alienation. What a relief. How refreshing. How wonderfully refreshing. Just listening to Jeremy talk about the stories of he and his friend, Steve van Hemert, and all their escapade—it's just the rebelliousness that is so healthy. And seeing this inspiration can re-ignite in all of us our moral instinctual power. We have the proof about who we are and we can finally make the great turnaround from this race toward our terminal alienation."

[52] Stefan Rössler: Watching this video and reading Tony's post and all the comments is just so inspiring! Yes, God, this is needed—someone to stand up and just say it like it is, without pretending that we've got it all together when we clearly don't.

[53] This is just so inspiring, so relieving, so refreshing, and so incredibly needed. Thank God for Jeremy and Steve and everyone at FIX THE WORLD, and for all the bravery and defiance it takes to overthrow the old, dark cave-world and bring about the glorious new world of understanding and sunshine ☀.

[54] Ales Flisar: This ignites all the engines in each one of us, IT SENDS US ON OUR WAY, let's go, let's crank this up big time. Yes we are all extremely upset and all that stuff but that does not matter anymore, what matters is to get this absolute love of humans out there for everyone to know about. This is what happens with the human condition explained, solved and made obsolete. WOOOOOHHHHHHOOOOOOO, massive thanks to Stevie.

[55] Frank Balamatsias: What a great post Tony, and timeless speech which will go down as one of the greatest! Incredible to hear of the wild high-spirited adventures of Stevie and Jeremy and how it ultimately led to the resolution of our crippled condition. Boy did we need someone to stand up against it, they're both Ned Kelly like in their defiance.

[56] Genevieve Salter: I love the significance of the flag, I love the defiant escapades, I love the boldness and beauty of Steve's truth-telling honesty leading to Jeremy's soul-alive that solved the human condition and lives and fights with every breath for the wonderful, truthful, all-loving world in sunshine that we're all at last coming home to. Yes, let's hold the fort til the cavalry arrives. Let's live for this with every breath as Jeremy does.

[57] John Nichol: Amazing to hear Jeremy talk about his and Steve's big bold adventures in cracking open the briefcase of bullshit 😄. It's such a refreshing contrast to the small lives most of us have lived under the duress of the human condition. Let's all take this inspiration and break free of our chains and end human suffering once and for all 💥.

[58] Claire Rickie: What fun Jeremy & Steve had in the 60s, so adventurous & humorously exposing of the artificial ways we all use to prop ourselves up—loved this video.

[59] Karen Boon: Absolutely wonderful 💛 💛

[60] Anna Rytsy: Love this!!!

[61] Brony FitzGerald: Yes, so deeply inspirational to help us all stand up against the silence and denial all around us and within us. We know our job, just support this understanding! Jeremy and Steve

putting an almighty stop to the destructive bullshit worldwide with Jeremy's compassionate truthful explanation of all of us extremely busted alienated humans! Just beyond all description inspirational!

[62]Monica Kodet: Absolutely out of this world! Jeremy never gives up on humanity, all the time with flawless logic busting us out of our small, strangled-by-the-human-condition worlds.

[63]Ari Akritidis: I could not love this video more. Just so, so special.

[64]Sally Edgar: The world needs this redeeming explanation so, so desperately, what a privilege to see Jeremy in action showing us how to get this information to everyone.

[65]Yvonne Hayes: I've already watched this video half a dozen times! And Tony's post is just as inspiring! **'Our job at FIX THE WORLD is to hold the fort until the cavalry of responsible establishment support arrives'**, as evidenced by those 2 magnificent commendations that the cavalry is getting ready to charge!!

[66]Lachlan Dunn: Gold indeed Tony.

[67]Sam Akritidis: Loved watching Jeremy's special commemoration and tribute to one of his best mates, the legendary larger-than-life Stevey van Hemert.

[68]Neil Kohler: YEEEHAAA, let's go and take over this (Denial Earth) town 🤠— Love it!

69 Nikoletta Akritidis: This is so, so, so beautiful. Love hearing about how precious Stevie was to Jeremy and how important their friendship was to be able to solve the human condition.

70 Pete from Ireland: Wow, that video of Jeremy's speech just made my April. How lucky we are, how clear our choice is, how strong we can now be.

71 Kevin Ryan: Oh Thank You!! This is like breathing in the freshest of air. WOW WOW WOW. Jeremy and Steve, thanks!!!! LGTFOOH!

www.ingramcontent.com/pod-product-compliance
Lightning Source LLC
LaVergne TN
LVHW052349100826
845147LV00012B/793

* 9 7 8 1 7 4 1 2 9 1 1 8 6 *